Simple Question, Unorthodox Answer.

Twinkle, Twinkle Little Star

How I wonder what you are.

A paper discussing the brightness and distance of the stars in the heavens.

Andrew Robert Chapman 10.9.2023

Edition: 29.05.2024

Before continuing I suggest that the reader spend a couple of minutes looking at the stars. You may never look at them the same way again after reading this paper.

I'll be using current scientific thinking and theories as well as mainstream (Wikipedia etc.) internet links. I heartily encourage and welcome any challenge and counter-argument to this thesis' final question.

Indeed, the first to answer me the question will receive a cash reward.

Contents

TWITTER Summary

Our Sun shines with a luminosity of $1L^1$. By the time its light reaches Earth it is already perceived as a much smaller, brighter and hotter disc than it truly is. By the time its light reaches Pluto, 6 billion km from the Sun, it already looks like any other star in Earth's night sky.

Our nearest star, Alpha Centauri (A) shines with a luminosity of 1.5L. and is 41.5 trillion km away from us.

Question: How can the light of a star, which is one and a half times brighter than our Sun, travel nearly 7000 times further (41.5 trillion divided by 6 billion) and still be perceived to be as bright and large in the Earth's night sky as our Sun would appear when viewed from Pluto?

---

[1] The **solar luminosity** ($L_\odot$) is a unit of radiant flux (power emitted in the form of photons) conventionally used by astronomers to measure the luminosity of stars, galaxies and other celestial objects in terms of the output of the Sun.

One nominal solar luminosity is defined by the International Astronomical Union to be 3.828×10^{26} W.[2]

Management Summary

Our Sun has a diameter of about 1.4 million km and is about 100 times bigger than the Earth (with a diameter of 13 thousand km). The Earth orbits the Sun at an average distance of 150 million km. At this distance the Sun happens to appear as large as our Moon, which is just 3500 km in diameter and orbits Earth at an average of 380 thousand km.

In Space then, just as on Earth, objects appear smaller the further they are away from the observer.

If we were living on the planet Mercury, which is closer to the Sun, the Sun would appear larger in the night sky. If we were to live further away, on Mars, for example, the Sun would appear to be smaller and, if we were looking at the Sun from the furthest (dwarf) planet in our solar system, Pluto, then the Sun would look like any other star in the Earth's night sky (see Figure 1)

Pluto is 6,000,000,000 (6 billion) km from the Sun.

Our Sun shines with a brightness of 1L. Our **nearest** Star, Alpha Centauri (A), shines with a brightness of 1.5L. It is 41,471,300,000,000 km (41.5 trillion km) away from Earth.

Question: How can the light of a star, which is one and a half times brighter than our Sun, travel nearly 7000 times further (41.5 trillion divided by 6 billion) and still be perceived to be as bright and large in the Earth's night sky as our Sun would appear if viewed from Pluto?

Logically and, more importantly, mathematically, if Alpha Centauri shines one-and-a-half times brighter than our sun, then it can only be can be one-and-a-half times further away[2] than the 6,000,000,000 (6 billion) km Pluto is from our Sun.

9,000,000,000 (9 billion) km.

If Alpha Centauri is 41,471,300,000,000 km (41.5 trillion km) away, then it must shine with a luminescence of around 7000L to make itself visible in our night sky.

Closer or brighter.

One or the other.

---

[2] Unfortunately for science and mathematics, one and a half times is very generous. As we shall read and discover later, due to the inverse-square law, the distance isn't even one and a half times.

Diagrams: NTS (Not To Scale)

Diagrams (see <ins>Figure 4</ins>) representing the Sun, planets and universe are NEVER, EVER even remotely to scale. (Technical) drawings are annotated with NTS (Not To Scale) to denote this.

As a practical exercise to demonstrate the futility of representing planetary bodies to scale for the reader, we shall try to actually draw a part of our Solar system on paper. We will use a scale of 1mm representing the equivalent of 1000km.

Grab a sheet of A3 paper or, if you have no A3, two sheets of A4 placed end to end will suffice.

We're going to draw the Earth and Moon (to scale).

Draw a 1.3cm (13mm) line near the edge of the paper. This represents the diameter of the Earth (13,000km). If you have a compass draw a circle, otherwise sketch a freehand circle or simply leave the line (-), plus (+) or asterisk (*) to represent the Earth's approximate circumference.

Our Moon (3,500km) will be a 0.35cm (3.5mm) line/circle but where must it be drawn? Well 380,000km (the Moon's orbit around the Earth), scales down to 38 cm. That is then on the second sheet of A4 paper or near to the opposite side of an A3 sheet.

If we were to draw the Sun using this scale it would be a disc with a diameter of 140cm (1.4metres) and it would be 150 (one hundred and fifty) *metres* away from our 1.3cm Earth!

150 metres would therefore represent one Astronomical Unit (AU) using this scale and, for those interested in sharing a pizza on Mars with Elon Musk in 2025[3] you would find Mars a further 75 metres from the Earth, 225 metres from the Sun. Mars would be 0.6 cm (60 mm) in diameter.

Remember: to date we've made it, in the late '60s and early '70s, a distance of 38 cm from the Earth and next up is nearly 200 times that distance; 7,500 cm, to Mars.

If we wanted to draw Pluto on this scale model we'd have to walk (run or maybe drive or cycle) a distance of 6,000 metres (6km) to plot a 2mm wide "circle" representing our solar system's outermost dwarf-planet.

If we wanted to accurately draw Alpha Centauri A using this scale we'd have to travel 41,471,300km! The Earth,

---

[3] People will go to Mars by 2025. "If things go according to plan, we should be able to launch people probably in 2024 with arrival in 2025," said Musk.

with a circumference of 40,075km, is, in fact, nowhere near big enough for our scale drawing. We would literally have to drive around the Earth 1,034 (one thousand and thirty four) times to represent the distance to scale.

Therefore all diagrams and illustrations of planets and stars which you see in books (and even on video) cannot ever be anything other than "NTS".

A quick "back of the envelope" calculation to check that the scale diagram is accurate one can calculate the angle between Earth, Moon and Sun. It should be approximately the same (remember the Moon almost perfectly blocks out our Sun during an eclipse).

By using elementary school mathematics (SOHCAHTOA) the radius of the Moon/Sun becomes the "Opposite" side of a right-angled triangle. The "Adjacent" side is the distance from Earth to the centre of the Moon/Sun. The angle formed is then the Tangent (TOA) of Opposite divided by Adjacent.

For the Moon we calculate: (3.5/2)/380 = 0.004605 (26° 38')

For the Sun we calculate: (1400/2)/15000 = 0.00466 (26° 73')

And bear in mind that after just "150 metres" (1 AU) our Sun (1400 mm) is effectively as big as our moon (3.5 mm).

Discourse: Closer or brighter

On a clear night we see many bright "pin-pricks" of light shining above us. We call them "the stars" and, during the day, their luminance is overwhelmed by the brightness of the light our own star; the Sun.

Current scientific theory states that our Sun is separated by about 93,000,000 miles (nearly 150,000,000 kilometres) from our planet the Earth and it is a good job that there exists so much distance between us and our life sustaining star.

Entertain me, if you would, by ignoring our knowledge of real-world physics for a paragraph or two. Let us disregard the Sun's heat and the blinding light and consider only distance.

If our Earth were right next to our Sun, it would, obviously, because of its vast size, completely fill our sky, from horizon to horizon. The heat and the intensity of its glare, which we are not considering for the moment, would, of course be, literally, world destroying. At least for that fraction of a fraction of a second the earth existed, before being vaporized into atoms and elements.

The brightness alone would be unimaginably intense but we can, indeed have, calculated its brightness, or luminosity, L.

Our Sun has been handily been assigned the value 1L (also shown as $L_\odot$) by Earth's astronomical scientists. It has a radius of nearly 700,000 km (about 109 times that of our Earth).

We all know, despite the distance between Sun and our planet, that the Sun is still so bright that it will permanently damage our eyes, should we look directly at it for anything more than a brief period. But its light and warmth are largely responsible for the continued existence of life on Earth.

When we do dare to look at the Sun, when it is dimmed behind a summer's cloud, for instance, we observe that it is a "bright round disc" which we can easily "cover" by holding a coin, or similar sized disc, with outstretched arm. In fact, we are gifted with the extraordinary coincidence that our Earth's single moon is of such a size (even with the moon illusion) and distance that it perfectly covers the sun during an eclipse. So we all know how relatively large the sun is, because, at night, we can look at the moon (which reflects the Sun's light back at us), with no fear of damaging our eyesight.

The Earth's moon has a diameter of about 3475km and orbits Earth at an average distance of about 380,000 km and, perceptively, this is (as near as exactly) close enough to block out the disc of the Sun which, as we've stated, is 1,400,000 km in diameter and shining its light at a distance of around 150,000,000 km from the Earth.

The point is: the massive Sun appears to be as big as our moon and this is directly attributable to the much larger distance between the Sun and Earth than the Moon and Earth.

The Sun's brightness/luminosity and its radiated heat are also, and most gratefully, massively reduced from those which exist on its surface and this again is directly attributable to the large distance between the Sun and Earth.

We can calculate these properties using "luminosity formulae" and more specifically, the "relationship to magnitude".

Luminosity is the intrinsic value of the light source (for (a practical) example, how bright a hand-held torch actually is) and the apparent magnitude is its brightness when observed from a distance. Your torch's light effectively, and obviously, becomes less bright, or dimmer, the greater the distance between it and the person observing it. Just as our Sun's light dims, the further we travel away from it.

Just as the light from a torch will be dimmer if, for example, it is foggy then the light from stars is also dimmed more due to space debris, gases and dust.

For the purposes of this paper we will **not** be considering this and always assume a perfectly, space-dust free universe; an assumption which ultimately aids any counter-argument to this paper's final question.

On Earth one could place a magnifying lens, in much the same way as a laser works, in front of a torch light. This lens would concentrate the rays of light emitted from the torch and bundle them together, so that the brightness is greater than it normally would be.

There is, as far as I am aware, no such mechanism of increasing the luminosity of a star and the light will remain (for all intent and purposes of this book) constant (although, when sun's die then the light they emit does vary, as in the case of an exploding supernova).

Simply put: The further away from the Sun a planet is, the smaller the disc of light from our Sun will be and the less heat and light (brightness) is received.

On the edge of our solar system, where the currently titled "dwarf planet", Pluto, is orbiting, at a distance of 3,700,000,000 miles (nearly 6,000,000,000 km) or Eris 6,400,000,000 miles (over 10,000,000,000 km) the Sun will not be much bigger or brighter than the stars we see in the Earth's night sky.

Hold that thought.

The Solar system consists of our Sun and its celestial bodies which orbit around it; the eight Planets, including Earth, and dwarf planets (including Pluto and Ceres and Eris) and asteroids, rocks, dust and gases. But mainly there is a whole lot of nothingness. In fact, statistically speaking, our Solar system, indeed space itself, is effectively nothingness with a miniscule clump of matter we call the Sun and a few even tinier bits of matter we call planets.

In our Earth's night sky several of what we call "stars" are, of course, our solar system planets, which are reflecting our Sun's light back to Earth and our eyes in much the same way as the moon reflects the Sun's light back to us.

These colloquially named "stars" (Mars, Venus, Jupiter etc.) are **not** the subject matter of this paper when the word "star" is used.

Each star we see in the sky is a faraway sun, more than likely with its own system of orbiting heavenly bodies (planets, asteroids etc.).

Light, travelling at the speed of, erm, light, from our Sun takes around 8 minutes to reach the Earth. The light which reaches our planet's surface stops here. Some of its light, which doesn't hit the planet, will be bent by our atmosphere. But, as previously mentioned, the Earth is a tiny, tiny speck of dust and pretty much all the Sun's light goes on to shine in all its glory into the unfathomable depths of the universe.

Light speed is 1,080,000,000 km/h and thus covers the 150,000,000 km to Earth in around about 8 minutes.

Distance divided by Velocity gives Time.

$D / V = T$

150,000,000 km/ 1,080,000,000 km/h=0.13888 hours

0,13888 x 60 (minutes in an hour)=8.3 minutes (0.3 of a minute is 20 seconds)

We are, when we look at the Sun from the Earth, in fact looking at what the Sun was like 8 minutes (and 20 seconds) ago.

The Sun's light obviously takes longer to reach the planets whose orbit is larger than the Earths and, if we were sitting at a café on Pluto, looking at the speck of light in the night sky, which is our Sun, 3,700,000,000 miles (nearly 6,000,000,000 km) away, it has taken the light 5 and a half hours to make the journey.

All calculations used in this paper are dealing with "average" orbits. The planets do not follow a perfectly circular orbit and this (along with axial tilt) is a part of the mechanism which gifts our Earth its seasons. But, for the purposes of this thesis, using the average orbital distance is in no way significant for its calculations and assumptions.

In a similar way you'll notice the figures and results are rounded. The paper states 8 minutes instead of the official 8 minutes and 17 seconds or 5 and a half hours instead of 5 hours 33 minutes and 20 seconds). Accuracy to the nth degree really isn't necessary when pitted against the vastness of our Solar system and the universe. And so, getting back to those other stars in our night sky, those suns of faraway solar-systems, how long has their light travelled to reach us?

The "*nearest*" sun to our solar system is Alpha Centauri and, coming in at a distance of 4.37 light-years, "near" is definitely a very subjective description.

Whereas the light from our Sun needs 8 minutes to reach the Earth the light from Alpha Centauri has taken 4.37 years! The speed of light is a constant: light travelling from our Sun is travelling exactly as fast as the light from any and every other star.

We won't consider the gravitational braking effects of black holes on light as there are no black holes between us and Alpha Centauri.

Light, travelling at 1,080,000,000 km/h manages to cover a distance of 25,920,000,000 km in a (24 hour) day. Lets call it a round 26,000,000,000 km. For a year we'll use 365 days and drop the extra leap-year day every four years.

Using these figures we calculate that light travels 9,490,000,000,000 km (that's near enough 9.5 trillion km) in an Earth year! That figure, multiplied by our 4.37 light years, comes to 41,471,300,000,000 km and represents the distance between our Sun and Alpha Centauri.

41,471,300,000,000 km (41.5 trillion km).

The figures are, literally, so astronomically large that (understandably) scientific notation (4.1471×10^{16} m) is used to represent these distances and, in fact, new units of measurement have been invented by astronomers (for example the <u>parsec</u>) in order to keep the reader of their scientific papers and theses going goggle-eyed when they try to count digits and commas.

Our fantastically large distance of 41,471,300,000,000 km, for example, is reduced to a feeble and paltry sounding 1.34 parsecs. But this paper will resolutely use the large numbers and familiar units of measure to explicitly impress upon the reader the "vastness" of the distances being discussed.

The reader must agree that all the stars in the Earth's night sky are, more or less, about the same size and brightness; certainly to within a small band of "discrepancy". There are no stars, for example, which are three-quarters, a half, a quarter or any other fraction of the size of the Sun they are all, more or less, pin-pricks of light and they are all, more or less, of the same brightness.

Alpha Centauri, our closest star (apart from the Sun) is (<u>when viewed through telescopes</u>) noticeably larger and brighter but, again, not "dominatingly" so. In other words: The light reaching the Earth from our "nearest" neighbouring star, Alpha Centauri, is pretty much the same as all the stars in our night sky.

We recall that, equally, these "specks" of light in Earth's night sky are also, approximately, as big and bright as our own Sun would be if we were to view it from Pluto, at the edge of our own Solar system.

This therefore means that the light emitted from Alpha Centauri, travelling a distance of 41,471,300,000,000 km is, relatively, about as bright as our Sun after its light has travelled just 6,000,000,000 km.

Let's be generous and nearly double the distance and state that our Sun, when viewed from a distance of 10,000,000,000 km, (instead of the actual 6,000,000,000 km) is as bright as Alpha Centauri is to us on Earth.

Even using this (generous) assumption it necessitates that Alpha Centauri, which is 4,147 (over four thousand!) times further away from us than our own Sun, logically be at least 4,147 times brighter and more luminous than our Sun.

It actually has to be much (much) more than that.

The "Inverse-square law"

Up till now the exact manner and mathematics in which light "dims" over distance has been avoided but luminosity actually "dims" according to the inverse-square law.

WIKIPEDIA has a section explaining the mathematics and even has a section where the light from the Sun and our planets are mentioned. Here's a copy of the relevant text from the section headed "Light and other electromagnetic radiation" (the bold italics are mine):

"The intensity (or luminescence or radiance) of light or other linear waves radiating from a point source (energy per unit of area perpendicular to the source) is inversely proportional to the square of the distance from the source, so an object (of the same size) twice as far away receives only one-quarter the energy (in the same time period).

For example, the intensity of radiation from the Sun is 9126 watts per square meter at the distance of Mercury (0.387 AU); but only 1367 watts per square meter at the distance of Earth (1 AU)—*an approximate threefold increase in distance results in an approximate ninefold decrease in intensity of radiation*."

In other words, one can't halve the (relative) distance of the star, as we have done, and arrive at a luminescence of 4,147 $L_\odot$

One has to take the square of this figure.

If Alpha Centauri were 20,000,000,000 km away it would be separated by 2 (radius) lengths (20,000,000,000 km divided by 10,000,000,000 km) and need to shine at 2 squared. 2 multiplied by itself is 4. So, at 20,000,000,000 km Alpha Centauri would need to shine at 4 $L_\odot$ in order to be as bright as our Sun (1 $L_\odot$) when viewed from Pluto (a (generously) rounded 10,000,000,000 km) away.

If Alpha Centauri were 30,000,000,000 km away it would be separated by 3 (radius) lengths (30,000,000,000 km divided by 10,000,000,000 km) and need to shine at 3 squared. 3 multiplied by itself is 9. At 30,000,000,000 km Alpha Centauri would need to shine at 9 $L_\odot$ in order to be as bright as our Sun when viewed from Pluto.

At 40,000,000,000 km it is separated by 4 radius lengths and need a luminescence of 16 $L_\odot$ (4 squared)

Separated by it's "true" 4,147 radius lengths Alpha Centauri would need a luminescence of 17,197,609 $L_\odot$ (4,147 multiplied by 4,147).

17 million $L_\odot$ (instead of 4 thousand $L_\odot$) but, unfortunately, WIKIPEDIA doesn't confirm this.

The brightest of the Alpha Centauri suns, Alpha Centauri A, is listed as having a luminosity of 1.5 $L_\odot$! That is not even twice as bright as our Sun. Alpha Centauri B "weighs in at 0.5$L_\odot$.

Let's again be "generous" and assume Alpha Centauri A shines with a luminosity of 4$L_\odot$.: More than twice the official 1.5 $L_\odot$.

If we were to place the star Alpha Centauri A directly next to our Sun and go to our viewing terrace 10,000,000,000 km away, we'd see our Sun as a pinprick of light and Alpha Centauri A shining four times brighter next to it.

Were we to go to a viewing terrace a further 10,000,000,000 km away Alpha Centauri A would be as bright as our Sun was at the first viewing terrace and our Sun (four times dimmer) next to it.

Having travelled a distance of just 20,000,000,000 km the light from Alpha Centauri A (and we've been very generous giving it a luminosity of 4 $L_\odot$.) would hardly be discernible from the other stars in the universe.

How can it be that the light from Alpha Centauri travels a further 41,451,300,000,000 km (41,471,300,000,000 km - 20,000,000,000 km) and is still (very) visible in the Earth's night sky?

As a reminder; these rough calculations don't take into account the light damping effects of space dust and gases, maybe even asteroids and planetoids, of which there are statistically bound to be some to interfere with the light on a 41 trillion kilometer trip taking over 4 years.

Furthermore the "back of the envelope"4100 figure is actually over 7000 (when calculated using the WIKIPEDIA figures (see the <u>Management Summary</u>)).

Perhaps the compensating factor is the sun's size?

But the size of a light source doesn't, of course, make any difference to the distance it travels. The brightness or luminosity is the sole deciding factor. Imagine your torch as big as a satellite dish, but still only as bright

But, before we even get into the mathematics and physics of that argument, WIKIPEDIA lists the size of Alpha Centauri A as 1.2 times that of the Sun. So, even if size were to matter, which it does not, Alpha Centauri doesn't have it anyway.

The question this thesis poses is: how can the light from a star, Alpha Centauri A, which is marginally bigger and marginally more luminous than our Sun and Alpha Centauri B which is marginally smaller and half as luminous as our Sun be seen in Earth's night sky when the light from our own Sun demonstrably fades and dims into the blackness of the universe long, long, long before it has traversed the massive distance of 41,471,300,000,000 km?

Obviously something in my thesis' assumptions doesn't add up. But what?

The only guesstimate in my assumption is that the Sun is a pinprick of light when viewed from 10,000,000,000 km away.

Is it really?

A quick browse of the internet turned up this picture: (see Figure 1)

This illustration only reinforces this paper's thesis (especially considering the fact that Pluto is 6,000,000,000 km from the Sun and this paper uses a (very generously) "rounded up" distance of 10,000,000,000 km).

In other words, our Sun really would look just like any other star in the night sky when viewed from a distance of 6,000,000,000 km (never mind 10,000,000,000 km).

This article, "How Does the Sun Appear on Other Planets?", is actually sub-titled, "From Pluto, it would be just another star.". The article also includes a picture of how colours and contrasts are lost over the ever increasing distance (see Figure 2)

So it would seem that this paper's presumption, that Sun is "just another star" when viewed from the (relatively small distance) of 6,000,000,000 km, is not just perfectly sound but also widely accepted common knowledge. That our Sun would seem like a normal star when viewed from Neptune or Pluto would, by no means, appear to be a groundbreaking idea or heretical to established scientific thinking.

How then can we see the light from other stars in our night sky? Stars which are orders of magnitude (the **nearest** being over 7000 times) further away from us than our Sun? We should not be able to see them, even with the aid of optical telescopes.

But see them we (obviously) do!

There are only two factors which can affect how we perceive the light emitted from all the visible stars in our Earth's night sky. From our "close" neighbour, Alpha Centauri thru all the others suns which are (much) further away.
Those two factors are distance and luminosity.
If Alpha Centauri (A or B) are not significantly brighter than our Sun then they have to be significantly closer than the official 41,471,300,000,000 km.
If the Alpha Centauri stars are 41,471,300,000,000 km away from the Earth, then they have to be significantly brighter. Around 49,000,000 times (7000 squared) brighter!

Closer or brighter.

Whilst Alpha Centauri may be familiar to Earth's dwellers living in the Southern Hemisphere, those in the Northern Hemisphere, including the author, are more familiar with, say, the Orion constellation. Randomly picking one of the stars in that very visible nightly constellation, Rigel, we can read in WIKIPEDIA, under the subheading „Physical Characteristics"
"Other calculations based on theoretical stellar evolutionary models of Rigel's atmosphere give luminosities anywhere between 83,000 L$\odot$ and 363,000 L$\odot$"
It won't make a jot of difference but let us be generous and take the upper estimate of 363,000 $L_\odot$. And assume (again generously) that it is ("only") 264 parsecs away (there are estimates that it is 360 parsecs away, but that won't do the science any favours as we've learned, the

19

further away a star is, the more luminosity is lost due to the inverse square law).

We have already calculated that Alpha Centauri needs to be shining at 49 million $L_\odot$ if it is 41,471,300,000,000 km away.

Alpha Centauri is 1.34 parsecs away. Rigel is therefore 197 times (264 parsecs/ 1.34 parsecs) further away than Alpha Centauri (around 8,169,846,100,000,000 km, 8 thousand trillion or 8 quadrillion kilometres).

As a reminder, if we were to be drawing Rigel on our NTS diagram, instead of 1,000 times around the Earth (as we would have had to have done with Alpha Centauri) we would have to circumnavigate Earth's equator 197,000 times for Rigel!

Applying the inverse-square law:

If we take the distance from Earth to Alpha Centauri as our new "radius", then we know that there are 197 radii to calculate.

197 squared is 38,809. However we need to factor in the luminescence of Alpha Centauri, which we calculated as 49 million.

49,000,000 $L_\odot$. multiplied by 38,809 gives us the required luminosity of Rigel IF it is to be seen by our eyes on Earth at a distance of 264 parsecs.

Rigel needs a luminosity of 1,901,641,000,000 $L_\odot$. That's nearly two trillion $L_\odot$!

Closer or brighter.

What Do I Know?

Effectively that which I was taught in school and learned afterwards from reading books on astronomy or watching documentaries and, as astronomy is not a subject of great personal interest to me I'd posit about as much as the average layman.

Nevertheless, taking a step back and looking at the world through a "child's eyes" can sometimes help remove the "blinds" from the eyes which blind the specialists who are so busy working on the forefront of their science that they pay no more heed to the simple, "settled" stuff.

But my question, at the start of this section was "What do I know?" not "What have I learned and been told?"

I "know" what I can experience and prove to myself. I "know" as much as you and, effectively, all the world's population. The same as all but a handful of (now) old (or dead) men who flew to the moon. But that's a story for another thesis/paper.

I can "see" that we have a large "light emitting disc" in our sky, which we call the Sun. This definitely generates heat and light (because both of which are reduced should a cloud pass in front of it). The Sun is about the same relative size as our moon. I haven't personally witnessed a full-eclipse to be able to state it is the exact same size, but when I look at the moon at night and compare it to the size of the sun the next day I'd say they're about the same size.

That's really about it! That's all I can claim that I know.

I have no idea how big those discs in the sky really are. How far away, or how hot or luminous. I have no idea if our Sun really is a massive, fiery, flame-spewing fireball 150,000,000 km away Those are things I "know" only because I have been told or taught them. The Sun could just as easily be a (much) smaller, not so fiery and hot and much closer source of light and heat.

Effectively, with just my uneducated "knowledge", I am unable to prove anything. I would need to know either the distance to the Sun or its luminosity to work out the other missing factors.

Dead or alive

I offer a monetary reward to any person who can disprove this thesis. The reward of €250 will stand until 24.2.2024 and will be awarded to the first person to write to me with an explanation to this e-mail address:

ARCinfo33@proton.me

Should the thesis be disproved[4] I will extend and publish this paper with the eMail's contents.

I will not speculate here about what the alternatives could be, I'll save that for yet another paper, but I will leave you with a clever, a prossibly fitting, quote:

"When you have eliminated the impossible, whatever remains, however improbable, must be the truth?"
— Arthur Conan Doyle, *The Sign of Four*

---

[4] Disprove stipulates using mainstream scientific theories alone. Those which are found in WIKIPEDIA: For example, if there were an entry stating that "the luminance of light remains constant after travelling a distance of 6 billion kilometres" then I'd accept that my paper is incorrect.

Flaws in my calculations will be equally accepted as valid.

Illustrations

What would the Sun look like from Jupiter or Pluto?

Mercury	Venus	Earth	Mars	Jupiter	Saturn	Uranus	Neptune	Pluto
1.3 degrees	42	30	20	6	3	1.5	0.9	75

Figure 1

Figure 2

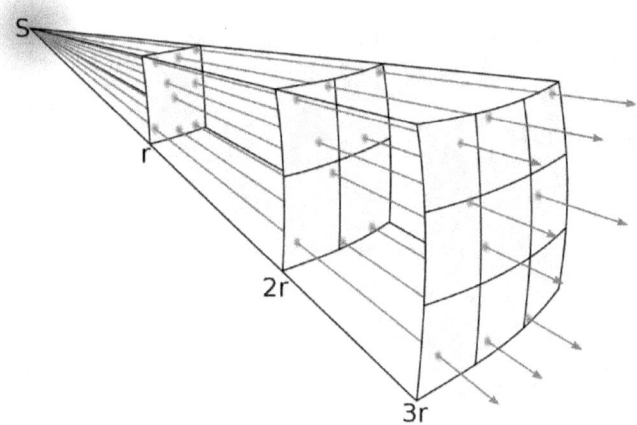

Figure 3 Luminosity degradation due to the inverse square law

Figure 4 Trust me. This is definitely NTS

Twinkle, twinkle, little star,
How I wonder what you are!
Up above the world so high,
Like a diamond in the sky.

When the blazing sun is gone,
When he nothing shines upon,
Then you show your little light,
Twinkle, twinkle, all the night.

Then the trav'ller in the dark,
Thanks you for your tiny spark,
He could not see which way to go,
If you did not twinkle so.

In the dark blue sky you keep,
And often thro' my curtains peep,
For you never shut your eye,
Till the sun is in the sky.

'Tis your bright and tiny spark,
Lights the trav'ller in the dark,
Tho' I know not what you are,
Twinkle, twinkle, little star.

by Jane Taylor (1783–1824) (my emphasis)

Epilogue

Interestingly enough, it was a TV documentary about the Pioneer space probes which set my mind off thinking about this paper's conundrum.

As the TV documentary's animation showed the space probes on the edge of our Solar system, our Sun was just another tiny speck of light, indistinguishable amongst the thousands of other stars ...

A PDF of this book/paper is located at:

https://www.mediafire.com/folder/ceownpycrxwxg/

There you will also find PDF files of Answers I have received and my Response to them. Thus keeping this publication focused and tight.

Other books by the Author

Please visit your favourite eBook retailer to discover other books by Andrew Robert Chapman:

z/OS Guide Series
ISBN-13: 979-8601937375:
COBOL: Optimised & Maintainable Application Programming
ISBN-13: 979-8602445213: JCL – STEP by STEP
ISBN-13: 979-8882582233: Practical z/OS HLASM
I
Songwriting Series
ISBN-13: 978-0463202951: F.A.T.E: ALPHAs & BETAs
ISBN-13: 978-0463492048: PackEis: Lyrics and BETAs
ISBN-13: 978-0463844366: WildßcreW: A Graphical Revelation
ISBN-13: 979-8354626465: G.A.S.C. and I
Lyrics and Poetry Series
ISBN-13: 978-0463277454: 2011 & earlier: Lyrics & Poems
ISBN-13: 978-0463393758: 2012: Lyrics & Poems
ISBN-13: 978-0463205938: 2013: Lyrics & Poems
ISBN-13: 978-1370272846: 2014: Lyrics & Poems
ISBN-13: 978-0463610473: 2015: Lyrics & Poems
ISBN-13: 978-0463923771: 2016: Lyrics & Poems
ISBN-13: 978-0463472866: 2017: Lyrics & Poems
ISBN-13: 978-0463173268: 2018: Lyrics & Poems
ISBN-13: 978-0463388174: 2019: Lyrics & Poems
ISBN-13: 979-8582846918: 2020: Lyrics & Poems
ISBN-13: 979-8832205168: 2021: Lyrics & Poems
ISBN-13: 979-8832205168: 2022: Lyrics & Poems
Corona Pandemic Series
ISBN-13: 979-8859838721: Saving Granny

www.ingramcontent.com/pod-product-compliance
Lightning Source LLC
Chambersburg PA
CBHW072239230526
45466CB00025B/2187